URANUS

MARYSA STORM

BLACK RABBIT BOOKS

Bolt Jr. is published by Black Rabbit Books
P.O. Box 227, Mankato, Minnesota, 56002.
www.blackrabbitbooks.com

Michael Sellner, designer and photo researcher

Names: Storm, Marysa, author.
Title: Uranus / by Marysa Storm.
Description: Mankato, Minnesota : Black Rabbit Books, [2023]
Series: Bolt Jr. space explorer | Includes bibliographical references and index. | Audience: Ages 6-8 | Audience: Grades K-1 | Summary: "Suit up! Uranus is a planet like no other, and it's time to explore it through a colorful design that features vibrant photography, engaging infographics, and carefully leveled text"– Provided by publisher.
Identifiers: LCCN 2021026236 (print) | LCCN 2021026237 (ebook) | ISBN 9781644665992 (paperback) | ISBN 9781623107789 (hardcover) | ISBN 9781623107864 (ebook)
Subjects: LCSH: Uranus (Planet)–Juvenile literature.
Classification: LCC QB681 .S796 2023 (print) | LCC QB681 (ebook) | DDC 523.47–dc23
LC record available at https://lccn.loc.gov/2021026236
LC ebook record available at https://lccn.loc.gov/2021026237

Image Credits

Alamy: Buradaki, 6–7; Sebastian Kaulitzki, 10–11; Dreamstime: Forplayday, 22–23; Getty Images: MARK GALICK/SCIENCE PHOTO LIBRARY, 12–13; iStock: dottedhippo, 10–11; Science Source: Julian Baum, 4–5; Mark Garlick, 1, 8–9, 10–11, 16–17, 18–19, 20–21; Monica Schroeder, 20–21; Shutterstock: Bikomins, 3, 24; Mopic, 14–15; NASA, 6–7; Pe3k, 4–5; Vadim Sadovski, Cover; Vectorfair.com, 14-15; Vector Tradition, 12–13

Contents

CHAPTER 1

Meet the Planet

It was 1977. Scientists sent a **probe** into space. Nine years later, it flew by Uranus. It was the first spacecraft to reach it. It took pictures. It sent back info.

probe: a machine used to collect information from outer space and send it back to Earth

COMPARING SIZES

Uranus

99,018 miles
(159,354 kilometers)
around

Big and Blue

Uranus is a big planet. It is a gas giant. That means it's made mostly of gas. The gas gives the planet a blue color.

Earth
24,901 miles (40,074 km) around

Features

made mostly from water, ice, and gas

cold and windy

CHAPTER 2

Its Place in Space

All of the solar system's planets circle the sun. While circling, they spin like **tops**. But not Uranus. It's tilted. It spins like a rolling ball.

top: a toy with a point on which it is made to spin

FACT

Long ago, a large object might have hit Uranus. It would have caused its tilt.

Far From the Sun

Uranus is an **outer planet**. It's far from the sun. Being far away makes it cold. Temps drop to -370 degrees Fahrenheit (-223 degrees Celsius).

outer planet: any of the solar system's planets that lie beyond the asteroid belt

Uranus in the Solar System

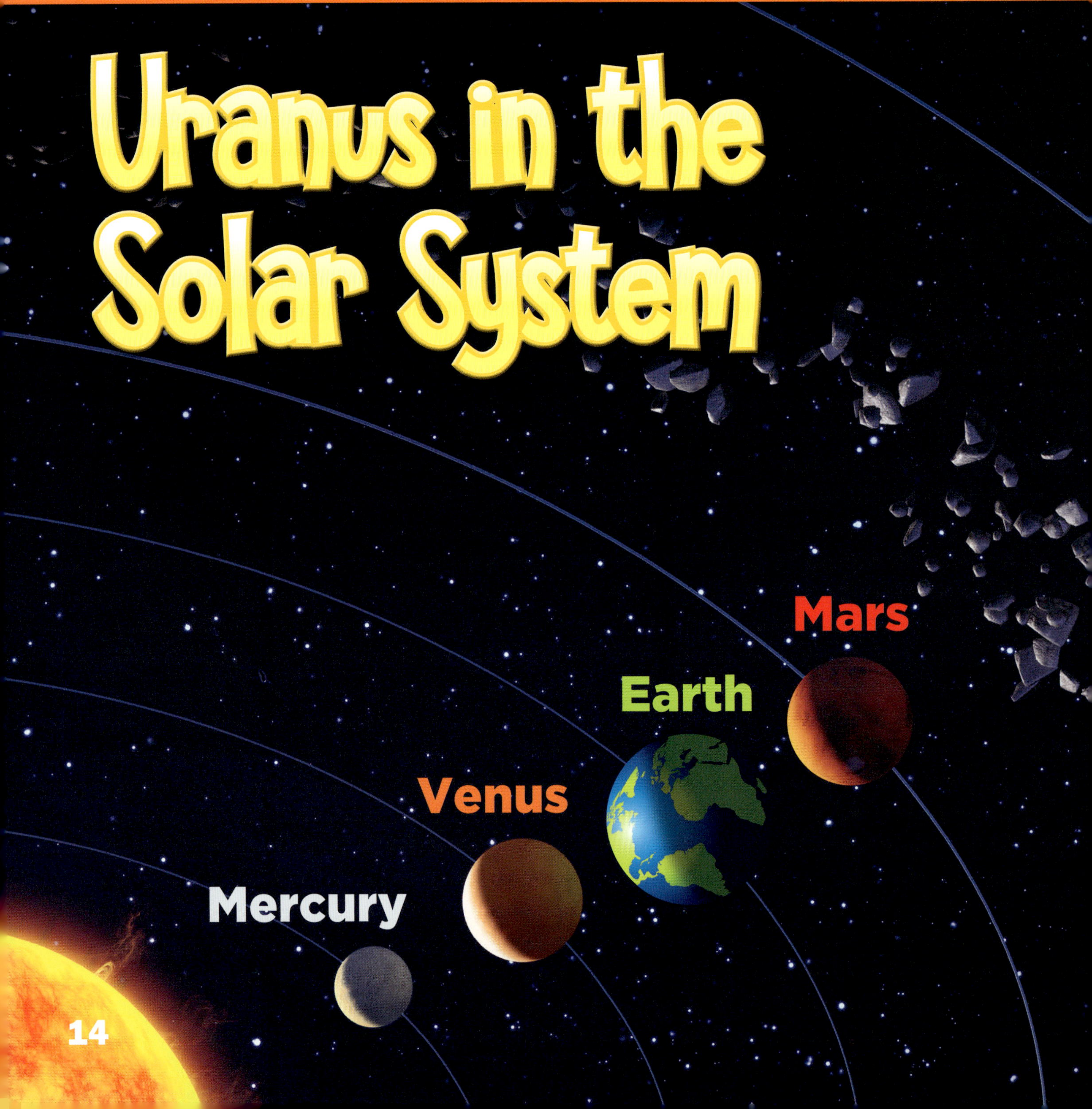

Neptune
Uranus
Saturn
Jupiter
asteroid
belt

CHAPTER 3

Explore!

Uranus is far from Earth. It's about 1.7 million miles (2.7 million km) away. It takes years to reach it. Only one craft has ever made the trip.

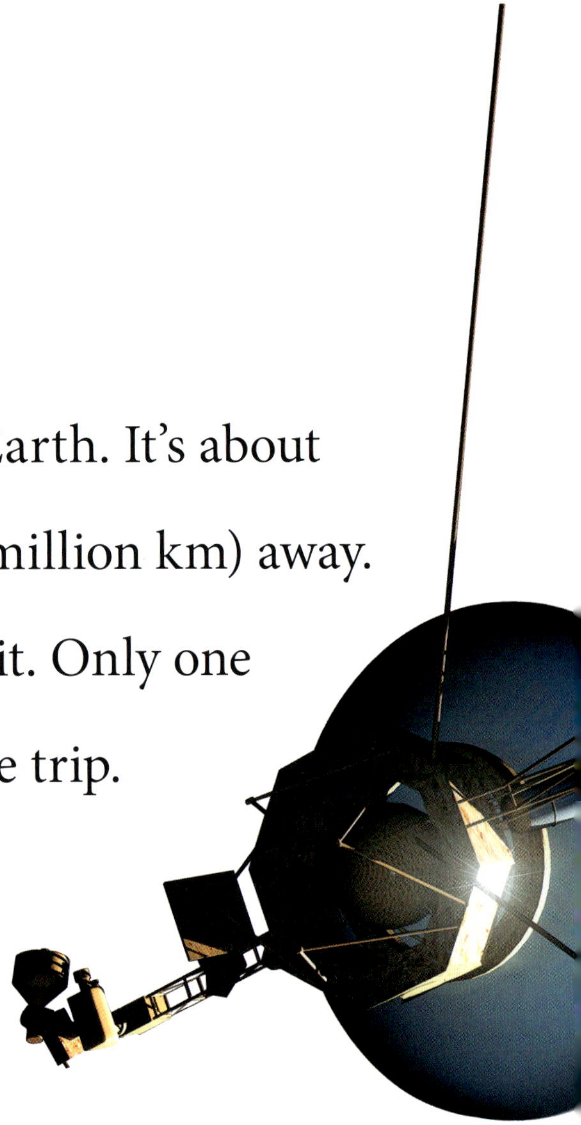

FACT

No more missions are planned.

The Future

For now, people study Uranus with telescopes. They want to learn more about its moons. They hope to find new rings.

number of moons

27 moons

Bonus Facts

It has 13 known rings.

One day lasts **17 hours.**

The planet was discovered in 1781.

Its color comes from methane.

methane: a gas that has no smell and can be burned for fuel

READ MORE/WEBSITES

Foxe, Steve. *Uranus.* Planets in Our Solar System. North Mankato, MN: Pebble, an imprint of Capstone, 2021.

Grack, Rachel. *Curious about Uranus.* Curious About. Mankato, MN: Amicus/Amicus Ink, 2022.

Loh-Hagan, Virginia. *Space.* Weird Science. Ann Arbor, MI: Cherry Lake Publishing Group, 2022.

Astronomy: Planet Uranus
www.ducksters.com/science/uranus.php

Mission to Uranus
kids.nationalgeographic.com/space/article/mission-to-uranus

GLOSSARY

methane (METH-eyn)—a gas that has no smell and can be burned for fuel

outer planet (OU-ter PLAN-it)—any of the solar system's planets that lie beyond the asteroid belt

probe (PROHB)—a machine used to collect information from outer space and send it back to Earth

top (TOP)—a toy with a point on which it is made to spin

INDEX